Muhammad Hassam Wajahat

Salt Precipitation during Gas Production

Muhammad Hassam Wajahat

Salt Precipitation during Gas Production

Numerical Modelling of Radial Flow

LAP LAMBERT Academic Publishing

Impressum / Imprint

Bibliografische Information der Deutschen Nationalbibliothek: Die Deutsche Nationalbibliothek verzeichnet diese Publikation in der Deutschen Nationalbibliografie; detaillierte bibliografische Daten sind im Internet über http://dnb.d-nb.de abrufbar.
Alle in diesem Buch genannten Marken und Produktnamen unterliegen warenzeichen-, marken- oder patentrechtlichem Schutz bzw. sind Warenzeichen oder eingetragene Warenzeichen der jeweiligen Inhaber. Die Wiedergabe von Marken, Produktnamen, Gebrauchsnamen, Handelsnamen, Warenbezeichnungen u.s.w. in diesem Werk berechtigt auch ohne besondere Kennzeichnung nicht zu der Annahme, dass solche Namen im Sinne der Warenzeichen- und Markenschutzgesetzgebung als frei zu betrachten wären und daher von jedermann benutzt werden dürften.

Bibliographic information published by the Deutsche Nationalbibliothek: The Deutsche Nationalbibliothek lists this publication in the Deutsche Nationalbibliografie; detailed bibliographic data are available in the Internet at http://dnb.d-nb.de.
Any brand names and product names mentioned in this book are subject to trademark, brand or patent protection and are trademarks or registered trademarks of their respective holders. The use of brand names, product names, common names, trade names, product descriptions etc. even without a particular marking in this work is in no way to be construed to mean that such names may be regarded as unrestricted in respect of trademark and brand protection legislation and could thus be used by anyone.

Coverbild / Cover image: www.ingimage.com

Verlag / Publisher:
LAP LAMBERT Academic Publishing
ist ein Imprint der / is a trademark of
OmniScriptum GmbH & Co. KG
Heinrich-Böcking-Str. 6-8, 66121 Saarbrücken, Deutschland / Germany
Email: info@lap-publishing.com

Herstellung: siehe letzte Seite /
Printed at: see last page
ISBN: 978-3-659-75442-5

Acknowledgements

First of all I would like to thank Allah who gave me the strength and determination to write this thesis and bring my studies to a completion. I would not have been able to complete my studies without the help and support of a lot of people during this journey of two and a half years. I would like to thank my thesis supervisor Professor Dr. Pacelli L.J.Zitha for giving me the opportunity to work on this project and by supporting and motivating me during the most stressful period of my life by being a fatherly figure always available with valuable help and advice. I would also like to thank my TNO supervisors Aris Twerda and Paul Egberts who continuously provided support and took time out from their busy schedules in solving my problems regarding TOUGH2 and providing their valuable views and feedback during my time of need. Furthermore, this thesis would not have reached completion without the help of Pascal de Smidt and the student councillor John Stals who understood my situation and provided their invaluable support and assistance. I would also like to thank my examination committee members including my supervisors and Professor Karl Heinz Wolf and Durgesh Kawale who took out their invaluable time to read my thesis and accepted to be a part of my examination committee.

I would like to thank my mother and sisters for their financial and moral support during my studies and for always being there for me during my difficult times of need. Lastly I would like to thank all my friends especially Arsene Levi Keya and Khalid Saleh who helped me out and provided their valuable assistance and outputs during the completion of this work.

Finally I would like to dedicate this work to my late father who passed on two weeks after I started my studies leaving me alone and helpless at a critical time which provided me the strength in completing this work for his sake.

Abstract

The decline in productivity in mature gas fields is often attributed to liquid loading but in reality it may be a result of salt deposition which can have similar consequences. The phenomenon of salt precipitation is gradually increasing and is becoming a serious issue in the North Sea. The North Sea operators are aware of the severe consequences of the issue of salt precipitation in the gas wells. During gas production the reservoir brine will evaporate leading to an increase of dissolved salt concentration. Once the solubility limit is exceeded salt precipitation occurs which can have a detrimental impact. Salt precipitation results in a decrease in porosity and permeability and consequently in a decrease of gas production and eventually in killing and finally abandonment of the well. Salt precipitation increases towards the end of the life time of gas fields. Currently water washes are used to remove and dissolve the salts but a good understanding is still lacking as the problem is complex.

In this study we model radial flow near the well bore, further enhanced to a layered inhomogeneity structure. Our approach consists of numerical simulations using TOUGH2 with the EWASG module by which we determine the rate and the location of the salt deposition from the near well bore region. Additionally we run numerous simulations by varying certain parameters to check the effects on the location of the salt and the sensitivity of our model as a base for further studies regarding this complex subject. This study shows that the location of salt precipitation is influenced by parameters such as a change in differential pressures and salinity content while the introduction of heterogeneity does not influence the location of salt deposition for this particular model.

Contents

List of Tables

List of Figures

Nomenclature

General Symbols

ρ	Density of the substance (Kg/m^3)
S	Saturation
φ	Porosity
x	Mass Fraction
u	Darcy Velocity (m/s)
Q	Volumeteric flow rate (m^3/s)
L	Flow Length (m)
A	Area perpendicular to L (m^2)
K	Conductivity (m/s)
Δh	Hydraulic Head (m)
R	Radius of well (m)
g	Acceleration of gravity (m/s^2)
λ	Mobility (Two Phase theory relitive permeabilities)
λ	Sorting Parameter (van Genuchten model)
P	Pressure (Pa)
κ	Permeability (m^2)
μ	Viscosity (Pa.s)
S_S	Solid salt saturation
P_c	Capillary Pressure (Pa)
M^κ	Mass accumalation Term of component κ (Kg/m^3)
F^κ	Flux of component κ ($Kg/(s.m^2)$)
r_p	Pore radius (m)
σ	Surface Tension
P_0	Strength Coefficient (Pa)
T	Temperature (degree)
θ	Normalized porosity
ω	Area ratio for tubes (Verma Pruess Bundle of Tubes Model)
R	Universal Gas constant (J/(mol K))
Φ_r	Reduced Porosity
τ or G	Fractional Length of pore bodies (Verma Pruess)

Superscripts and subscripts

i	Phase component
κ	Particular component in that phase (component index)
ri	Relitive permeability of the component
a	Active phase
β	Phase index (liquid, gas, solid)
L	Liquid phase
G	Gas phase
1	Water Component, H_2O
2	Salt Component, NaCl
3	Gas Component, NCG
S	Solid Phase
lr	Irreducible liquid saturation
ls	Liquiid saturation at which the capillary pressure vanishes
gr	Irreducible gas saturation
nw	Non wetting
w	wetting
cap	capillary
bsat	Brine saturation
sw	Salt Contnet in well
* & ^	Already embedded in TOUGH2 codes in the van Genuchten Models

Relitive Permeabilities and Capillary Pressures

κ_{rl}	Relitive permeability of the liquid phase
κ_{rg}	Relitive Permeability of the gas phase

Abbreveations

EWASG	TOUGH2 module consisting of water, soild and gas
EOS	Equation of State
NCG	Non condasable gas
NaCl	Sodium chloride also referred to as halite.
NK	Number of mass components
NPH	Number of phases
VPL	Vapour Pressure lowering Effect
PI	Productivity Index

1. INTRODUCTION

The decline in productivity in mature gas fields is often attributed to liquid loading but in reality it may be a result of salt deposition which can have similar consequences. The problem of salt plugging in gas wells has been well documented in Germany and USA (Kleinitz et. al.).The phenomenon of salt precipitation is gradually increasing and is becoming a serious issue in the North Sea. The North Sea operators are aware of the severe consequences of the issue of salt precipitation in the gas wells.

The production wells in gas reservoirs often experience a rapid performance decline with the progress of recovery. In many situations this behaviour has been attributed to the halite scale in the near well bore region of the reservoir. Salt precipitation can have a detrimental effect on the gas producing well as the presence of halite can reduce its porosity, leading to a decrease in the permeability which could eventually lead to the killing and abandonment of the well. Currently, salt precipitation is reduced through washing techniques and inhibitors.

In order to obtain a better productivity control a better understanding of the salt precipitation phenomena is required which is currently lacking in literature due to complexity. If understood correctly there may be a possibility of reducing and even in certain cases eliminating the down-hole fresh water washes. Currently it is difficult to predict the location of salt deposition in the near well bore region and this study deals in making a simple radial flow model which can help predict the location of the salt deposition as a first step after which further studies can be conducted in order to reduce the problem.

The purpose of this thesis is to investigate the rate and more importantly the location of the salt deposition near the radial symmetric model. We model the problem using the simulator TOUGH2. TOUGH is based on the classical multiphase theory completed with relations that describe the possible effects of water evaporation and salt precipitation which can later help in predicting the decrease in permeability and porosity to give a better understanding and a direction in order to tackle the problem (Pruess et. al. TOUGH2 users guide). A simple model is created which is further enhanced for comparative studies and some sensitive parameters are investigated. This study is based on TOUGH2 and certain results are matched with DuMu^x (Paul Egberts presentation at TNO) at selected conditions for comparative study purposes as shown in Appendix B.

The following chapters deal more with this study. **Chapter 2** gives a general background of salt precipitation with the related work by various authors regarding its formation, complexity and the impact it can have on the gas wells and productivity. In **chapter 3** we describe our physical model and the main assumptions it depends upon. We then formulate the corresponding transport equations and the equations which are embedded in TOUGH2. Next we describe the conditions and enhance our simple model in a slightly more complicated layered reservoir in which we introduce heterogeneity in our model in order to investigate its impact on the location of salt deposition. Furthermore, we change various parameters, investigating their impact on the location salt deposition. In **chapter 4** we present our simulation results including pressure, saturation profiles, salt precipitation profiles and the gas productivity rates which is followed by the discussion of these results. In the final **chapter 5** the main conclusions and recommendations of this study are presented as a base for future work regarding the deposition of salt near the well bore. In the last portion is the **appendix** which has extra details regarding this thesis.

2. BACKGROUND

Salt deposition can occur when the solubility product of the dissolved ions exceeds a limit and this can take place either due to evaporation or the dissolution of the rock minerals. Evaporation can result in a reduction in water saturation which can in turn increase the salt ion concentration. In gas wells the evaporation occurs due to gas expansion near the well bore region.

In their study, Kleinitz et al. (2001) showed that the salt can accumulate in the well bore, perforation zone as well as in the reservoir. Salt deposits are often observed in the well bore and the tubing which can then be cleaned by periodic water wash operations. Furthermore, the authors discuss entrainment of reservoir water and halite precipitation in depleted gas wells. In their work Dietzel et al. (1998) is mentioned who in his work describes the expansion of water resulting from production of gas as a consequence of a decrease in reservoir pressure. Dietzel believed that the expansion leads to an increase in water volume and hence the water saturation above the connate water level thus mobilizing the water.

Literature has shown that there are two main reasons which could possibly result in the precipitation of salt. The first of these mechanisms is evaporation of the water from formation brine into the producing gas which in turn results in an increase in the brine salinity. Secondly changes in pressure and temperature can reduce the solubility of the salt in the brine. Due to this the brine can become salt saturated so the salt precipitates out. Literature also suggests that the decrease in pressure is a more significant factor hence it is the most likely cause of water vaporization resulting in salt deposition. There is a possibility that the pressure gradient in the reservoir can result in non-uniform drying rates resulting in localized salt deposition.

It has further been documented in the literature that gas production from a high pressure/ high temperature reservoir which contains brine shows plugging after a certain amount of gas has been produced. Place and Smith (1984) observed that a well was completely plugged after 1Bcf of production and it required a water wash stimulation treatment to restore its productivity. Other cases of deposition of scale in the well bore tubing in a high pressure and high temperature zone have also been reported e.g. in Jasinski et al. (1997). In all these cases the scale deposition has always been observed with a rapid reduction in pressure which has caused evaporation and the subsequent super saturation of the salt.

In their experiments on Berea sandstone by dry methane gas injection (Zuluaga and Monsalve 2003) showed that the sodium chloride salt concentrations increased above their respective solubility limits and caused the dissolved salt to precipitate. Their results showed that salt deposition as caused a reduction in the permeability of the rock core from 14 to 21%. In their work on water vaporization the authors identified two vaporization periods namely a constant drying period followed by a falling rate period.

A later study by Van Dorp et al. (2009) showed that evaporation and salt deposition in laboratory sand pack is has an important role to play under radial flow conditions. The authors carried out gas injection experiments in the conical sand packs under radial flow conditions and studied the evaporation of brine. With the aid of X-ray computed tomography they showed that the salt deposition takes place in the region near the producing end. In their work the authors hypothesized that the main cause for the greater salt deposition in the narrow producing end is caused by the higher pressure drops occurring due to radial flow conditions.

Further cases from literature which strengthen the idea that evaporation results in the salt deposition are from the works of Tang and Etzion (2003) who developed an empirical relation for the evaporation of water from a wetted surface. The model is a function of relative humidity and of the wind velocity over the water body surface. According to the model, a higher gas velocity will result in higher evaporation rates. For a radially flowing reservoir the gas velocity decreases as the distance from the well increases. This phenomenon results in the theory that salt precipitation due to evaporation of reservoir water during production of gas predominantly occurs near the well bore region where the gas velocity is the highest. In their works Zuluaga, Munoz and Obando (2001) performed an experimental study on vaporization induced halite precipitation in porous media and proved that vaporization increases with flow rate and decreases with the salinity in the brine.

Gas is initially saturated with water vapour at reservoir conditions. As soon as the production begins, the well pressure decreases near the well bore causing an increase in the solubility of water in the gas phase (Morin and Montel 1995; Kamath and Laroche 2000). Water starts to vaporize near the well bore where the pressure is low and the vaporization front moves into the formation. With the vaporization of the water the concentration of dissolved salts start to increase which could lead to the precipitation of the minerals. The impact of water vaporization can be severe as the precipitation of minerals can reduce the porosity, absolute permeability and ultimately kill the well leading to its abandonment.

3. MODELLING

TOUGH2 is a numerical simulator for non-isothermal flows of multi component, multiphase fluids in one, two and three dimensional porous and fractured media (TOUGH2 manual, Pruess et al., 1999). It is used in several applications including geothermal reservoir engineering, nuclear waste studies, environmental remediation and flow and transport in variably saturated media and aquifers. The current version includes several property modules offering enhanced process modelling capabilities like reservoir well bore flow, precipitation and dissolution effects and multiphase diffusion. A major disadvantage of TOUGH2 for oil and gas industries is that the industries demand user defined thermodynamic properties while TOUGH2 handles the thermodynamic data directly through the implemented equation of state (EOS) modules in a code (Lorenz and Muller, 2003).

3.1   EWASG (water, salt, gas)

Battistelli et al. (1997) developed the EWASG (water, salt, gas) module which was identified as a possible module to simulate halite precipitation aquifers (Lorenz and Muller 2003).This module was developed in order to model geothermal reservoirs with saline fluids and non-condensable gas (NCG). Several choices are available as a NCG including H_2, CO_2, CH_4, N_2 and air. The main assumptions and characteristics are:

1) Three mass components are considered; water, sodium chloride and a NCG
2) Water can be present in the liquid and gas phase
3) The salt component may be present in the liquid phase or it may have been precipitated to form a solid salt phase.
4) Solid salt is the only active mineral phase and is treated differently from the fluid phases. It is assumed to be immobile and its relative permeability is 0.
5) From the mass balances on salt in fluid and solid phases the volume fraction of the precipitated salt in the original pore space is calculated which is termed as "solid saturation" (S_s). The remaining space is available for fluid phases. (Section 2.2 and appendix for further details).
6) The gas phase cannot contain any trace of NaCl.
7) All relevant thermo physical properties are evaluated and the correlations used can be easily modified.
8) The dependence of brine enthalpy, density, viscosity and vapour pressure on the salt concentration is accounted for.

9) Vapour pressure lowering which is caused by the suction pressure effects is included in this model.

10) Transport of mass components occurs by molecular diffusion in the gas phase for the NCG and the liquid phase for all three components. The modelling of the hydrodynamic dispersion has not been included as yet.

11) It is assumed that all three phases (solid, liquid and gas) are in local chemical and thermal equilibrium and no chemical reactions take place other than the interphase mass transfer.

12) TOUGH2 and the EWASG module already incorporate several permeability porosity correlations e.g.Verma and Pruess (1988) embedded in its code. (See appendix C for more details).

13) For a system of three mass components according to the local thermodynamic equilibrium over the three phases, four independent variables or parameters are required in order to solve them. In the system of three coexisting phases, there is a possibility of seven combinations; three single phase conditions (solid, liquid, gas), three two phase conditions (solid-liquid, solid-gas, liquid-solid) and a three phase condition (solid-liquid-gas). EWASG module can handle six of these combinations excluding the single solid salt phase. The mass balance of the salt, water and NCG components along with the heat balance are set and solved in TOUGH2 by using the Newton Raphson iteration method.

14) Vapour pressure lowering effect due to the salt content in the brine

3.2 Physical Model

In order to tackle the problem of salt deposition near the well bore we decided to consider a 1D radial symmetric model. A homogenous model was built to investigate the salt deposition which was further enhanced to see the impact of heterogeneity by 2D radial flow (Layered Model). We designed a simple model focusing on well productivity and the impact of gas flow through the reservoir.

The model can be observed in figures 2 and 3 in which the first is our homogenous model while the second is the enhanced heterogeneous one. It is imperative to realize that we have modelled our problem over a grid size of 540 cells which is discussed in the grid refinement section.

Initially the model was developed for a 60 cell grid to examine the effect but after various simulations we decided that the optimum number of grid cells in order to investigate the effect would be 540. All our simulations are done and presented on this grid size. The total height of our model is selected as 24m while the lateral length of the model is 20m. This excludes the radius of the well which is 0.1m and accounts for a total lateral length of 20.1m.

In the layered model, we keep most of the conditions the same; the total height is again 24m. This is divided in three different layers each at a height of 8m for consistency purposes. The middle layer has a higher permeability than the top and bottom layers which are assumed to be identical in every aspect. All other conditions are kept constant in all the layers which are also identical with the original homogenous model. Each layer in the layered model is then further divided in four numerical layers decided by the mesh selection. This is further discussed appendix A. Furthermore, the simulations are conducted with an initial salt mass fraction of 0.26 over a period of four years.

3.2.1 Assumptions
We assumed the following:

1) The precipitated salt is immobile.
2) The initial water saturation has maximum halite concentration.

3) We consider isothermal conditions and hence pressure changes which may affect the temperature changes are disregarded.
4) There is no halite present in the gas phase (Place and Smith, 1984)
5) The well is designed explicitly in the model. The productivity index approach is not required as the cells are taken sufficiently small near the well allowing for an explicit modelling of the well. To ensure constant pressure boundary condition the first cell which is considered as the well and the last grid cell relating to the outer boundary are given a huge volume 10^{50}. This assumption is done in order to ensure that the thermodynamic conditions do not change at all from fluid or heat exchange.
6) A radial symmetric grid is chosen. Excluding the first cell (the well) which is relatively large the subsequent cells starting very small keep on increasing in size. This was followed in the initial 60 grid model. However, our final simulations are done on a 540 grid blocks and in order to get a similar geometry we do not change the size of the first and last cells but other cells are divided in 9 equivalent sections in order to keep them consistent.
7) The well is initially fully saturated with gas and there is no salt present in the associated grid block to avoid any sort of liquid flow which can cause water influx and thus bring an inaccuracy in the results.

3.2.2 Transport Equations

Mass Balance

The general mass balance equation for each component in two phase flow is described by the following equation:

$$\frac{d\rho_\beta S_\beta \varphi x_\beta^k}{dt} + \nabla\left(\rho_\beta \varphi x_\beta^k u_\beta\right) = 0 \tag{2.1}$$

Here the subscript β is the phase component and the superscript k represents the particular component in that phase, ρ, S, φ, x, u are the density, saturation, porosity mass fraction and the Darcy velocity respectively.

The above equation assumes that both the saturations and the mass fractions of the gas, liquid and the salt components sum up to 1. In the following summation k can be any of the three components.

$$\sum S_\beta = 1 \qquad \sum x_\beta^k = 1$$

The Darcy Law

The Darcy Law describes the flow of a fluid through a porous medium and was founded by Henry Darcy in 1856 and it states:

$$Q = AK\frac{\Delta h}{L}_L \tag{2.2}$$

Here $Q, A, K, \Delta h, L$ are volumetric flow rate, flow area perpendicular to L, volumetric conductivity, hydraulic head and the flow length.

Equation (2.2) was Darcy's Law for single phase flow which can be extended to two phase flow (2.3) using the relative permeability κ_{ri} as a scaling factor for phase β:

$$u_\beta = -\lambda_\beta \left(\nabla P_\beta - \rho_\beta g \nabla x\right)_L \tag{2.3}$$

Here λ, P are the mobility and pressure respectively

Equation (2.4) incorporates the relative permeability in equation (2.3)

$$\lambda_\beta = \frac{\kappa \kappa_{r\beta}}{\mu_\beta}$$ (2.4)

Here $\kappa. \kappa_\beta. \mu_\beta$ are defined as the absolute permeability, the relative permeability and the viscosity of the component.

Mass Balance in EWASG

In TOUGH2 the balance equations for a system of mass components (NK) is distributed in phases (NPH). (Pruess, 1991a). The general form for the mass accumulation terms $(\kappa = 1, NK)$ is given by the following equation:

$$M^k = \varphi \sum_{\beta=1}^{NPH} S_\beta \rho_\beta \, x_\beta^k$$ (2.5)

Here NPH is defined as the phase index and can be 1,2,3 based on liquid, solid and gas respectively.

The mass flux term is the sum over all the phases and is given by:

$$F^k = \sum_{\beta=1}^{NPH} x_\beta^k F_\beta$$ (2.6)

Individual phase fluxes are given by the multiphase version of the Darcy Law and equations (2.3) and (2.4) can be substituted which results in:

$$F_\beta = -\kappa \frac{\kappa_{r\beta}}{\mu_\beta} \rho_\beta (\nabla P_\beta - \rho_\beta g)$$ (2.7)

In the EWASG module NK=3, NPH=3, κ=1,2,3,4 which are water, NaCl, NCG and the respective heat components respectively, β=1,2,3 which indicates the gas, liquid and the salt phases respectively. It has already been assumed that the gas phase contains no NaCl and the solid salt phase is immobile (EWASG assumptions section 3.1.1) the accumulation and the mass flux terms for the salt component (κ=2) can be written by the following equations:

$$M^{(2)} = \varphi S_s \rho_s + \varphi S_L \rho_L x_{L^{(2)}}$$ (2.8)

$$F^{(2)} = -\kappa \frac{\kappa_{rL}}{\mu_L} \rho_L x_{L^{(2)}} (\nabla P_L - \rho_L g)$$ (2.9)

Here S_s is defined to be the solid saturation and it is the fraction of the pore volume which has been occupied by the solid salt.

Other Equations used

Several equations are used in the model including the relative permeability, capillary pressure and the vapour pressure effects. These are described in the following section:

Relative Permeability

As mentioned before, relative permeability is an important parameter which is used in the Darcy Law for multiphase flow. It is a function of the saturation of the respective phase. Furthermore, it depends on the saturations of the other phases and is a function of rock properties. It must be measured as a function of phase saturation for the given porous medium (Rossen, 2012).

The TOUGH2 codes already incorporate a number of relationships and in our model we make use of the van Genuchten- Mualem relationship.

$$\kappa_{rl} = \sqrt{S^*} \{1 - (1 - [S^*]^{1/\lambda})^\lambda\}^2$$ (2.10)

The gas relative permeability can be chosen in one of the following two forms depending on the conditions and the second form (2.12) is due to Corey (1954).

$$\kappa_{rg} = 1 - \kappa_{rl}$$ (2.11)

$$\kappa_{rg} = (1 - \hat{S})^2 (1 - \hat{S}^2)$$ (2.12)

Here $\kappa_{rl}, \kappa_{rg}, \lambda$ are defined as the relative permeability of liquid, the relative permeability of gas and the sorting parameter.

The parameters $\hat{s}, s^*$ are determined by the following relationships which are already incorporated in the TOUGH code.

$$S^* = \frac{(S_l - S_{lr})}{(S_{ls} - S_{lr})}$$ (2.13)

$$\hat{S} = \frac{(S_l - S_{lr})}{(1 - S_{lr} - S_{gr})}$$ (2.14)

Here $S_l, S_{lr}, S_{ls}, S_{gr}$ are defined as liquid saturation, irreducible liquid saturation, liquid saturation at which the capillary pressure vanishes and the irreducible gas saturation.

Capillary Pressure

The next equation that we define in our model is the relation of the capillary pressure function. Capillary pressure is an important parameter for the description of multiphase flow through porous media. The simplest formulation of the capillary pressure relates the fluid pressures in different phases and is defined as the difference between the non-wetting and the wetting phase as given by equation (2.15).

$$P_c = P_{nw} - P_w$$ (2.15)

However in the TOUGH2 code the capillary pressure is defined differently from the normal definition as is defined as the difference between the wetting phase and the non -wetting phase. This is the reason why equation 2.18 has a negative sign to balance the equations with the TOUGH2 code.

Where P_c is defined as the capillary pressure where as P_{nw} and P_w are the pressures at the non-wetting and the wetting phases.

Young- Laplace's equation (Christiansen,2008) was used to relate the capillary pressure difference between the wetting and the non-wetting phases by relating it to the radius of the curvature of the fluids assuming a spherical face R and surface tension σ.

$$P_c = \frac{2\sigma}{R} = \frac{2\cos\theta}{r_p}$$ (2.16)

Where r_p is the pore radius.

According to the Leverett-J Function the pore throat radius defined in the above equation can be expressed in terms of porosity and permeability to calculate the average pore radius as shown in (2.16).

$$r_p = \sqrt{\frac{8\kappa}{\varphi}}$$

(2.17)

TOUGH2 has many capillary pressure relationships relating with the liquid saturation embedded in it and we make use of the Van Genuchten (1980) formulation which is a useful relationship between the saturation and the capillary pressure and is given by:

$$P_c = -P_0([S^*]^{\frac{-1}{\lambda}} - 1)^{1-\lambda}$$

(2.18)

Here P_0 is a strength coefficient and S^* can be given by equation (2.13).

Vapour Pressure Lowering Effect (VPL)

The vapour pressure of water in a porous medium at a certain given temperature usually tends to be smaller than above a flat surface of bulk water. This is called the vapour pressure lowering and is caused by the capillary and vapour adsorbing effects. Collectively they are known as the vapour suction effects. (Pruess and O'Sullivan, 1992).

VPL has been incorporated in the TOUGH2 codes using the Kelvin Equation. We make use of these equations in our model (Batistelli et. al) and they are defined by the following:

$$P^{(l)}(T, S_{la}, x_{l2}) = f_{VPL}(T, S_{la}, x_{l2}) P_{bsat}(T, x_{l2})$$

(2.19)

Where f_{VPL} is defined by the following:

$$f_{VPL} = \exp\left[\frac{W^{(1)} P_{cap} S_{la}}{\rho_l(P_l, T, x_{l(2)}) R(T + 273.15)}\right]$$

(2.20)

The capillary pressure term in the VPL effect is evaluated by considering the active saturation of the liquid phase which is donated by S_{la}. The active saturations of the phases are defined as follows:

$$S_{\beta a} = \frac{S_\beta}{(1 - S_s)} \qquad\qquad (2.21)$$

Where $\beta = L, G$ and the components $S_{la} + S_{ga} = 1$

Permeability Change

The impact of porosity change on formation permeability is a very complex phenomenon and is one which has not been understood that well. Laboratory experiments have proved that small reductions in porosity from chemical precipitation have caused large reductions in permeability. (Vaughan,1987). This phenomenon was explained by the convergent- divergent nature of the natural pore channels where the pore throats can become clogged by precipitation while the disconnected void spaces remain in the pore bodies (Verma and Pruess,1988).

The EWASG module incorporates several of these permeability porosity relationships which bring about a change in permeability $\frac{k}{k_0}$ on relative change in the active flow porosity. The simplest model which can capture the converging-diverging nature of natural pore channels consists of alternating segments of capillary tubes with larger and smaller radii. This is shown in figure 1. This figure shows the bundle of tubes in series model in which the permeability is reduced to 0 at a finite porosity known as the critical porosity.

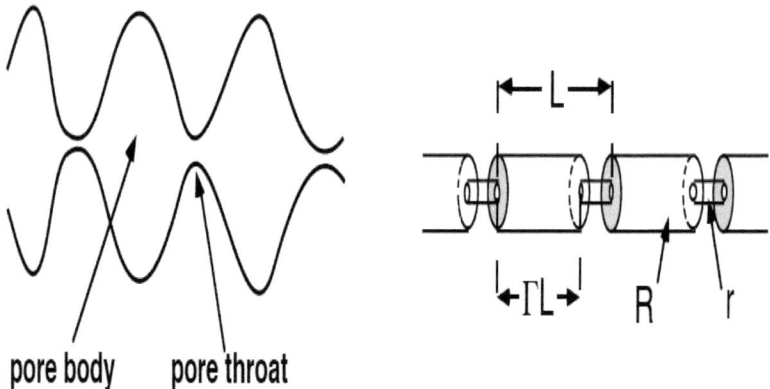

Figure 1: Model for the converging diverging pore channels (Verma and Pruess, 1988)

From the bundle of tubes model Verma and Pruess (1988) developed the following relationship which is embedded in the module and thus incorporated in our model.

$$\frac{\kappa}{\kappa_0} = \theta^2 \frac{1 - \Gamma + \Gamma/\omega^2}{1 - \Gamma + \Gamma[\theta/(\theta + \omega - 1)]^2} \qquad (2.22)$$

Here

$$\theta = \frac{1 - S_s - \Phi_r}{1 - \Phi_r} \qquad (2.23)$$

θ depends on the fraction $1 - S_s$ of the original pore space which is remaining in the fluid. It also depends on the parameter Φ_r which denotes the fraction of the original porosity at which the permeability is reduced to 0. Γ is the fractional length of the pore bodies and the parameter ω is given by the equation:

$$\omega = 1 + \frac{1/\Gamma}{1/\Phi_r - 1} \qquad (2.24)$$

3.3   Numerical Simulations

3.3.1 Simulation Models

In order to investigate the salt deposition phenomena we first model a 1D homogenous reservoir which is shown in figure 2.

Homogenous Reservoir Model (Base Case)

The homogenous model is our base case, which is used predict the location of the salt deposit. Simulations with this model indicate a certain distance from the well bore where the salt deposits. Having observed this, we vary and change our respective parameters to see the impact that they have on the production rate and the location of the salt precipitation. Some of the parameters which we investigated in this study included the effect of the initial liquid saturation, the effect of changing the bottom hole pressure and the effect of changing the salinity in the brine. The results of these cases are discussed in chapter 4.

Figure 2: Homogenous Reservoir Model (Base Case)

Heterogeneous (Layered model)

The layered model is an extension of the base case. The main reason for the development of the layered model is to see the impact of heterogeneity in a reservoir on salt precipitation. We focus mainly on the three different zones the top layer (T), the middle high permeability layer (M) and the bottom layer (B) in our simulations. A small section from each of these layers is taken to discuss the possible impact of salt depositions. In our simulations we focus on T4, M1 and B1. The reason why we chose 12 layers for our layered model was to increase the numerical accuracy of the simulations in relation to the mesh size which is further discussed appendix A.

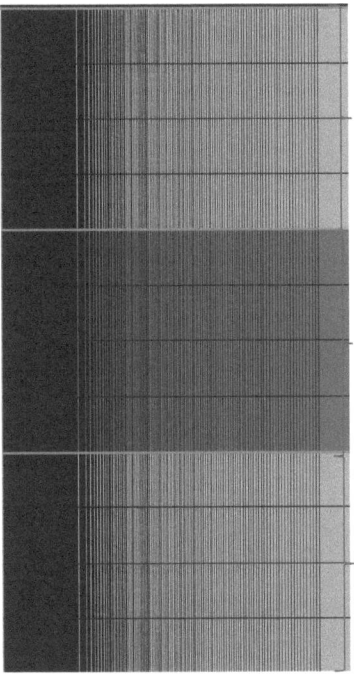

The top layer of the reservoir denoted as T. The top most layer is T1 and the bottom most one is T4.

The middle layer of the reservoir denoted as M. The top most layer is M1 and the bottom most one is M4.It has a higher permeability.

The bottom layer of the reservoir denoted as B. The top most layer is B1 and the bottom most one is B4.

Figure 3: Layered Model (Heterogeneous Reservoir)

3.3.2 Effect of Grid Refinement

Initially we designed a simple radial flow 1D model based on 60 grid cells and ran a simulation for a period of four years in order to investigate the impact on the location of the salt deposited. We took different snap shots at specific time periods and the salt saturation profile and the gas productivity curve are shown in figures 4 and 5.However, this grid size seemed too coarse and we repeated our simulated results by increasing the grid size.

We used different grid sizing and came to the conclusion that the best grid size for our simulations is the grid with 540 cells. The reason to choose this particular grid is that it is fairly close to the converging point as can be seen in figure 6. Furthermore, simulation run time was acceptable for this grid.

Another point of interest is that the effect of grid sizing has limited impact on the location of the salt deposited as shown in figure 7, however it does play a role in the rate of gas production as shown in figure 6.Salt deposition seems to occur earlier and the well blocks in a maximum of two years which is shown in all simulations.. A possible explanation for this could be the assumption that we are keeping a certain number of grid blocks equal in order to make it comparable with our first simulation so that we can follow the radial flow geometry.

The 60 cell seems too coarse and in this thesis all our simulations in the "results and discussion" chapter are done using this specific grid size 540 and for a time span of 4 years even though we observe that the salt has precipitated at about 2 years.

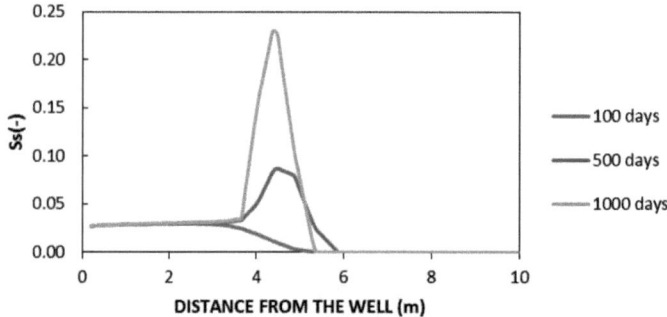

Figure 4: Initial Simulation salt profile at a 60 grid mesh

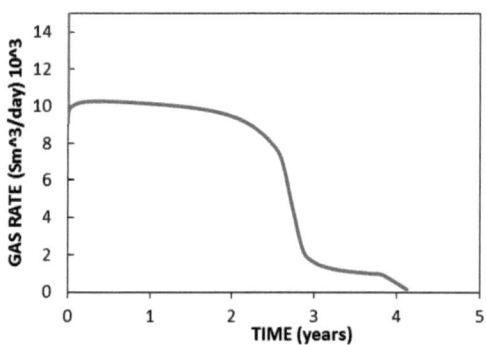

Figure 5: Gas Production Curve at a 60 grid mesh

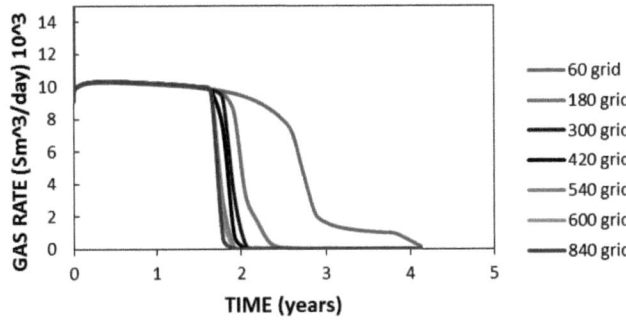

Figure 6: Effect of grid refinement

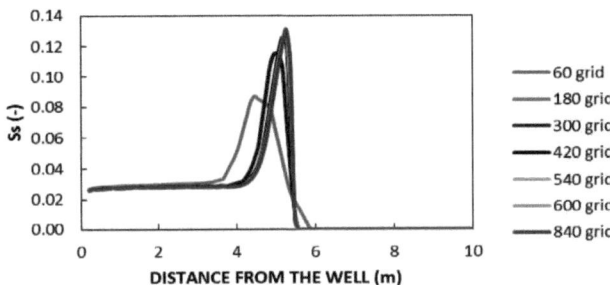

Figure 7: Impact of grid refinement on the location of salt at 500 days

3.3.3   Initial conditions, simulation parameters and model geometrical parameters

The purpose of the model is to investigate the location of salt deposition. Our results are then also compared to DuMu^x to see the impact of precipitation on another simulator. The model's dimensions, material properties and parameters are taken from real data for a well which had been collected. It has been modelled with values that can be observed in a typical reservoir. In both the homogenous model as well as the layered one all conditions are kept constant except the permeability which varies for the layered reservoir. Pressures at boundaries are fixed by ensuring huge volume

25

expansion of 10^{50} ensuring that the thermo dynamical conditions do not change. Our chosen values for the given parameters for the homogenous model are shown in table:

Table 1: General Parameters and values used in the model

Material Properties		
Name	Symbol (units)	Value
Height	h (m)	24
Density	ρ (kg/m^3)	2690
Porosity	φ (-)	0.11
Permeability**	κ (m^2)	2.23x10^{-14}
Well pressure*	P(Pa)	8x10^6
Reservoir pressure	P(Pa)	1.1x10^7
Temperature	T(θ)	145
Initial liquid saturation*	L$_S$ (-)	0.2
Gas saturation	G$_s$ (-)	0.8
Salinity in brine*	X$_s$(-)	0.26
Gas saturation in Well	G$_{sw}$(-)	1
Salt content in well	X$_{sw}$(-)	0
Radius of well	R(m)	0.1
Length of reservoir	L(m)	20

*Material properties that were changed from the homogenous case to investigate their impact on salt precipitation.

**The only parameter which is changed while developing the layered model.

As mentioned in section 3.2.2 we made use of the Van Genuchten relationships for our relative permeability and capillary pressure curves. The parameters that were used in these curves as well as for the Bundle and Tube models are given in tables:

Table 2: Relative permeability parameters (Van Genuchten Model)

Relative Permeability Curves ***(Van Genuchten-Mualem Model)		
Name	Symbol(Units)	Value
Sorting Parameter	λ (-)	0.6
Irreducible liquid saturation	S_{lr}(-)	0.045
Liquid saturation at which the capillary pressure vanishes	S_{ls}(-)	1
Irreducible gas saturation	S_{gr}(-)	0.001

Table 3: Capillary Pressure Curves parameters (Van Genuchten Function)

Capillary Pressure Curves ***(Van Genuchten Function)		
Name	Symbol(Units)	Value
Sorting Parameter	λ (-)	0.45
Irreducible liquid saturation	S_{lr}(-)	0.04
Strength Coefficient	1/Po(Pa)	4.0×10^{-5}
Maximum pressure	Pmax(Pa)	1×10^{9}
Liquid saturation at which the capillary pressure vanishes	S_{ls}(-)	1

*** The values in these tables are taken from literature and from TNO project where they were already tested.

In the above tables it is important to realize that the irreducible liquid saturation value for the capillary pressure curve is taken at a lower value than the relative permeability curve. In the original Van Genuchten derivation 1980 the irreducible saturations was the same in both the functions which resulted in the capillary pressure being infinity. This value is unphysical as it implies that the radius of the capillary meniscus goes to zero as the liquid phase becomes immobile. In reality no special capillary pressure effects are expected when the liquid becomes immobile. In order to get a more sensible and accurate results a smaller value of irreducible liquid saturation is used in capillary pressure curves in contrast to the relative permeability curves.

Table 4: Bundle of Tubes model Used

Bundle of Tubes in Series Model (Permeability-Porosity relationship)		
Name	Symbol(Units)	Value
Critical Reduced porosity ***	Φ_r (-)	0.7
Fractional Length of Pore Bodies****	G or Γ (-)	0.8

*** The value is taken from Literature TNO Report

****This is obtained from Literature from the paper of Verma and Pruess 1988.

3.3.4 Simulation Cases

Once our initial simulation was done we varied parameters like initial liquid saturation, differential pressure salt content in brine (base case) and permeability ratio (layered model) to investigate their impact on the location and the rate of salt deposition which are discussed in the results section. The purpose of these experiments is to obtain a better understanding of parameters that significantly influence salt precipitation, location and time scales and lay a basis for future studies once the effects are known. Our tests included the following cases shown in below:

Table 5: Cases examined on the base case

Homogenous Reservoir (Base Case)			
Case	Base Case	Experiment 1	Experiment 2
Initial liquid saturation	0.20	0.19	0.21
Differential Pressure	30 bar	20 bar	50 bar
Salt content in the brine	0.26	0.27	-

Table 6: Cases examined on the layered model

Heterogeneous Reservoir (Layered Model)				
Case	Base Case	Experiment 1	Experiment 2	Experiment 3
Permeability factor (M)*	κ (m^2)**	2κ (m^2)	5κ (m^2)	7.5κ (m^2)
Impact of gravity	9.81	No gravity	-	-

*(M) is the permeability of the middle layer is increased by a particular factor from a certain ratio.

** κ refers to the permeability based on the homogenous model 2.23×10^{-14} m^2. The scaling multiple from 2.0-7.5 is done on this value. The original value of κ for our middle layer in the layered model is 2κ hence 4.46×10^{-14} m^2.

4. RESULTS AND DISCUSSION

4.1 Base Case

We ran our simulations for a period of four years and took various snap shots within that time interval and came to the conclusion that our simulation ends around 700 days. From the salt saturation profiles in figure 8 we observe that the salt precipitates at a distance ranging between 4-5.5m. It goes to its maximum point of about 0.30 at this distance .Another observation which can be made from this graph is that the curve at 500 days superimposes the curve at 700 days on which basis we conclude that the salt precipitates around this time and the well starts to become blocked. For this reason we compared most of our simulations at the time of 500 days in order to observe the phenomenon before the blocking takes place.

The gas productivity curve verses time in figure 9 further strengthens the claim that the salt precipitates gradually over time and we observe a sharp decline around 500-700 days (1.7-2 years). Initially the gas starts to reduce slowly up till a period of 1.5 years which is exactly what we see in the salt profile that salt deposits forming uniformly up to 4m from the well after which there is a sharp rise in the salt precipitation resulting in the sharp decline of productivity.

In figures 10 and 11 we plot the saturation and pressure profiles of our simulation. The saturation curve shows that at a distance of around 4-5m from the well the gas saturation decreases and the corresponding liquid saturation increases as the gas is absorbing the water and due to this precipitation is forming at this distance which relates well with the observations of figure 8 where the salt saturation was observed.

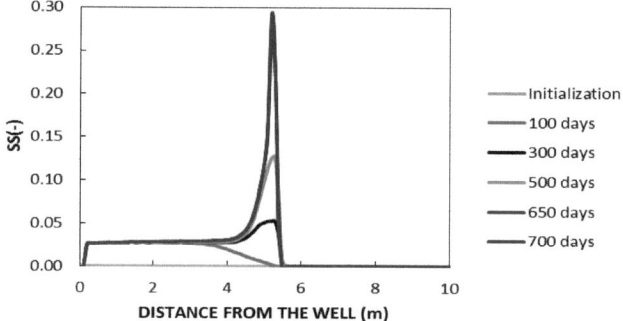

Figure 8: Salt saturation profile for the base case at different time intervals

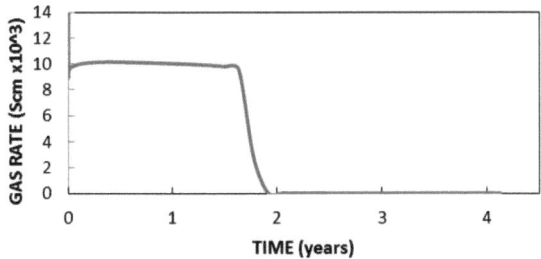

Figure 9: Gas productivity rate over the simulation period

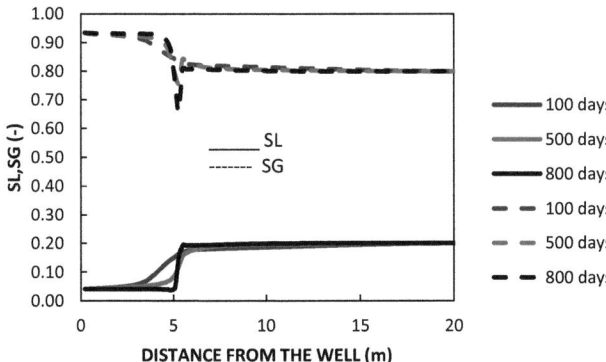

Figure 10: Liquid and gas saturation profiles for the base case at specific times

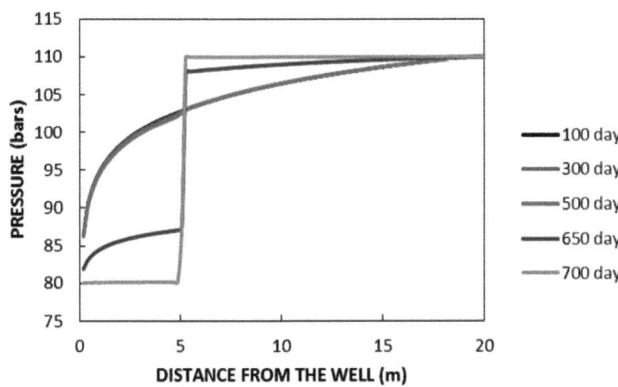

Figure 11: Pressure profile curves at various time

Effect of changing the initial liquid saturation

The effect of initial liquid saturation on the results is investigated by changing it from 0.19 to 0.21. As can be seen from figures 13 and 14 we see that the initial liquid saturation does not influence the location of the salt deposition. More or less the salt profile follows the same pattern as it did in the base case, gradually increasing with time. However, we note that at earlier time intervals salt saturation is sensitive to the initial liquid saturation. This is evident in figure 13 where we observe that at a period of 300 days the amount of salt saturated for the higher initial liquid saturation 0.21, is starting at a higher value of 0.03 and going to a maximum of 0.06 as well as precipitating closer to the well. Simulation ran with a lower initial saturation has a starting solid saturation of 0.02 going to maximum of 0.05 and depositing further to the well. We also observe that at later times of 500 days this effect is not visible anymore as shown in figure 14 and the initial liquid saturation is not a sensitive parameter for the location of salt deposit from the well.

Figure 12 further supports the corresponding salt saturation curves as it is observed that with an increase in the initial liquid saturation the gas productivity decreases at a faster rate. This can be attributed to a faster evaporation rate resulting in greater amounts of salt precipitation and the blocking of the gas reservoir. A higher liquid saturation means that the gas absorbs more liquid resulting in faster evaporation leading to salt precipitation at earlier times. We observe this clearly as the gas rate with liquid saturation of 0.21 becomes 0 well before 2 years and that of 0.19 longer. Furthermore, the gradient of the 0.21 curve is much steeper implying that salt precipitation is occurring faster as gas productivity is declining sharply in comparison to the case containing less water content.

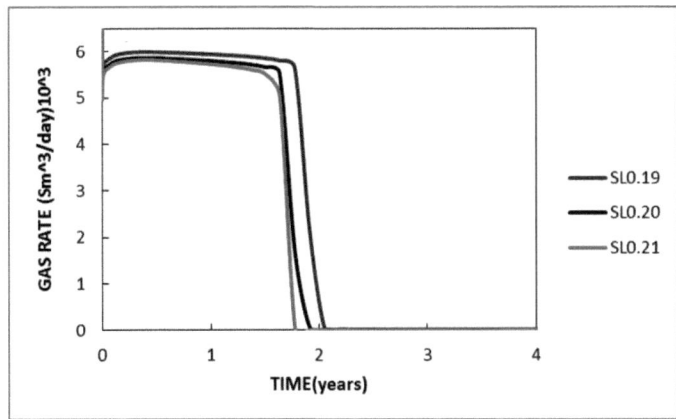

Figure 12: Comparison of the gas productivity rates by changing the initial liquid saturation

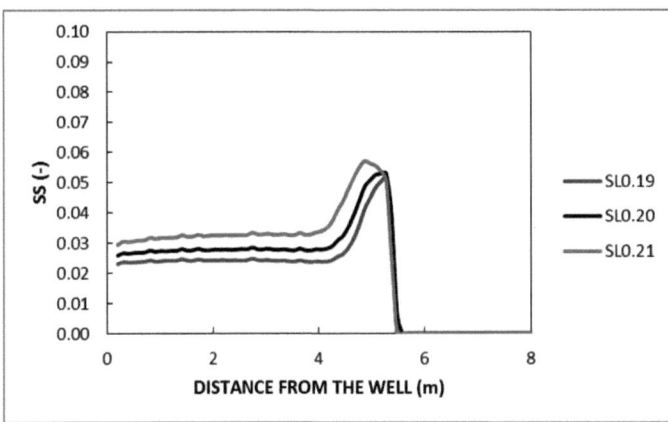

Figure 13: Initial liquid saturation is a sensitive parameter at earlier time intervals of 300 days

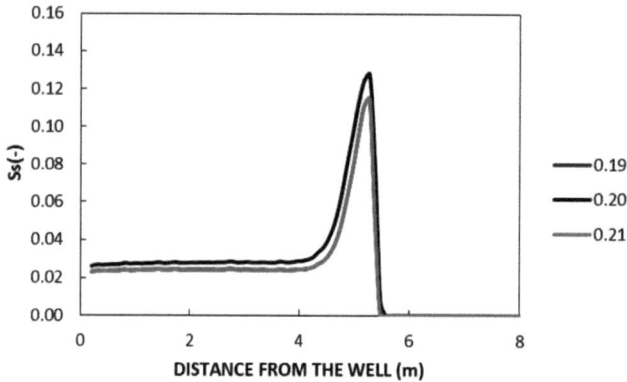

Figure 14: Initial liquid saturation salt profile is no longer sensitive at a later time of 500 days

Effect of changing the salinity content in the brine

The original simulation showed that salt deposits at a distance of 4.5-6m from the well at a salinity content of 0.26 However, changing it to 0.27 results in deposition of salt at a location further away from the well around 6.5-8m as shown in figure 15. Furthermore, in the same figure we also observe that the relative peak at higher salinity content is significantly lower than the corresponding lower salinity one. At a time interval of 700 days the simulation for 0.26 has reached completion and the well has been blocked completely which is not the case in salinity 0.27 where the Ss only reaches a value of about 0.10. Our simulations show that the model is sensitive to the amount of salt content in the brine. A small increase in the salinity content changes the location of the salt deposited considerably.

Fig 16 shows the respective gas production rates strengthening the influence of the effect of salinity on the location and the rates of salt precipitation. We observe that in the original case the gas rate becomes 0 after nearly 2 years while a small change in increasing the salinity results in the productivity being zero after a longer time 3.5 years. It is known from the works of Morin and Montel that an increase in salinity reduces the vaporization. Hence on basis of their work and our simulations we come to the conclusion that a slight change in the salinity content in the brine influences both the rate as well as the location of the salt deposition from the well and the model is sensitive to a change in this parameter.

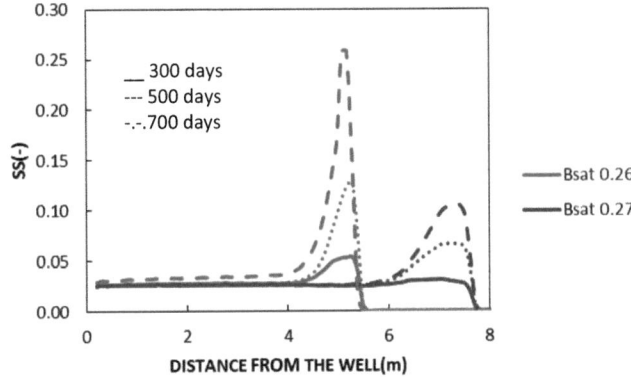

Figure 15: Effect of changing the salinity and its impact on the location of salt at specific time intervals

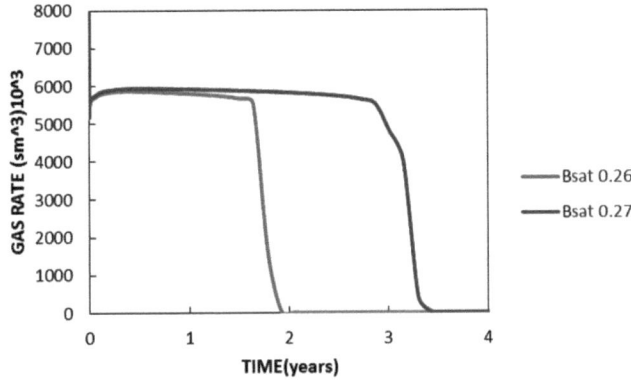

Figure 16: Impact of changing the salinity of brine on the productivity of the gas

Effect of Changing the Differential Pressure

From figure 17 it can be observed that at a lower differential pressure the salting phenomena takes place at a very fast rate. If we compare all our simulation cases we observe that initially salt takes time to accumulate increasing slowly before a sharp rise in the 4-5.5m range. However, this is not the case at lower differential pressures where the salt precipitations increases sharply right at the beginning of the simulation and keeps on rising. We observe in figure 17 that the blue curve for differential pressure 20 bars is at a higher value of 0.05 and keeps on rising steeply while pressure differentials at 30 and 50 bars the salt saturation profiles are uniform for a considerable distance at a value of 0.03. We also observe that at a lower pressure differential the salt deposited is much closer towards the well region as salt deposits around 1.7-2.5m at differential pressure of 20 bars. Likewise at a higher differential pressure of 50 bars the location of the salt is much further away from the well as it is deposited in the range of 7.5-8.5m.

Changing the bottom hole pressure has a significant impact on the productivity of the well. At a lower differential pressure the gas production is significantly lower and becomes 0 much faster in comparison to a higher differential pressure as shown in figure 18. We see that with a pressure differential of 20 bars the gas productivity becomes 0 after 1.2 years which is well explained by the corresponding saturation profile in which we observe the sharp peak at 500 days at a Ss value of 0.27 implying that the salt precipitates at a very fast rate at lower differential pressures. The steep gradient in the gas productivity curve further enhances this idea. Furthermore, as the differential pressure increases towards 50 bars it is observed that the gas productivity becomes 0 at around 2 years by which time the gas well is completely blocked. Our results show that another crucial parameter for determining the location of salt from the well is the influence of the pressure difference. Varying the bottom hole pressure has a significant impact on both the location as well as the rate of salt deposition

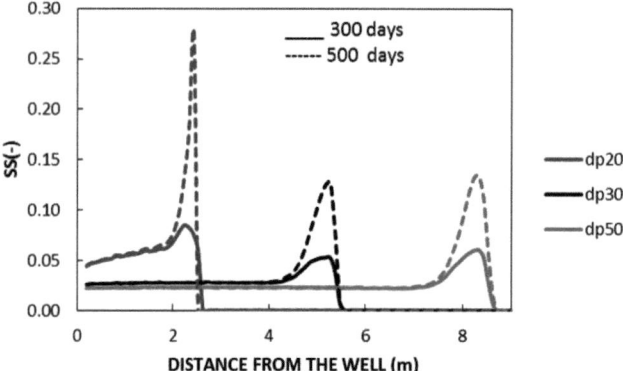

Figure 17: Effect of changing the differential pressure on the location of the precipitation of salt

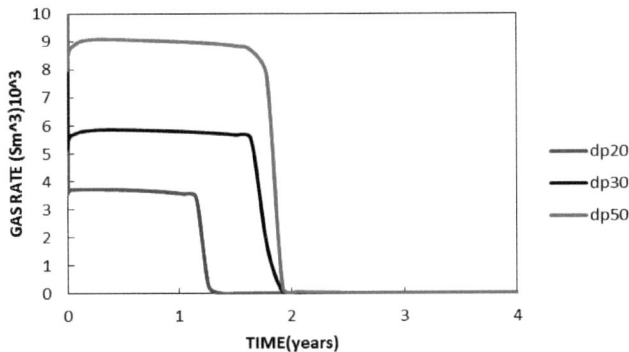

Figure 18: Impact of changing the differential pressure on the productivity of the gas

4.2 Layered Model

Location of salt Deposits in respective layers

As discussed in chapter 3 we introduced heterogeneity in our model by adding a layer of higher permeability in the middle to make a layered model. Keeping all conditions constant and equal in the top and bottom layers we observed that there was a difference in the amount of salt deposited which is seen in figures 19-21. Overall the salt deposits in the same manner as the base case but from figures 19-21 it is seen that a greater amount of salt is precipitated in the bottom region of the reservoir as compared to the top at all three time intervals. If we compare the interval at 500 days we see that the top layer has a Ss of about 0.17 while the bottom layer has a value of around 0.20.This immediately gives an idea that salt deposition is dependent upon the vertical position of the layer. Furthermore, we observe that the overall location of the salt is not dependent upon the heterogeneity of the reservoir. In figure 20 we have doubled the permeability but the salt deposition follows a similar pattern compared to the lower permeability region and the initial base case. We observe that in all three figures the salt is deposited at the same location from the well ranging between 4.5-6m. This phenomenon can be observed visually also when we see the vector profile of the gas being produced and can see the difference in the salt being produced in respective layers in figures 22 and 23.

However, we do notice one difference in the top and the bottom layers that with time the graphs are not smooth which is not evident in the high permeability region. We think a possible explanation could be the difference in heterogeneity as a higher permeability in the middle layer blocks the well faster and there is no permeability change in the lower permeability layers. The flow pattern changes from high permeability to lower permeability once the well has been blocked which may result in the abnormality of the graph. This is seen in the blue curves in figures 19 and 21 where there appears a kink in the production profiles for both the lower permeability regions.

We discussed that the introduction of heterogeneity in the model resulted in a difference in the amount of salt deposited between the top and the bottom region. This phenomenon is showed in figure 24. We find that the gas production rate declines at a faster rate in the bottom layer in comparison to the top layer. Both the curves start at around the same point at about 6000 m³ but around 1.5 years the gas production rate in the bottom layer has almost reached 0 while the top is still higher at 2000 m³. The top finally reaches 0 around 2.3 years by which time of the bottom layer of the reservoir has already been plugged. This implies that salt precipitation occurs faster in the bottom region as compared to the top relating fairly well with the respective salt saturation profiles of each individual layer.

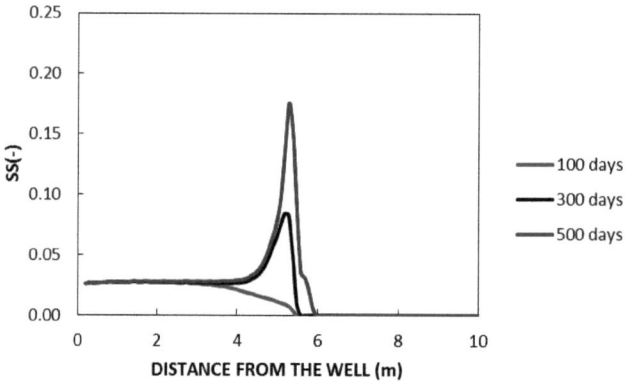

Figure 19: Salt saturation profile for the low permeability top layer in the layered reservoir at different time snapshots

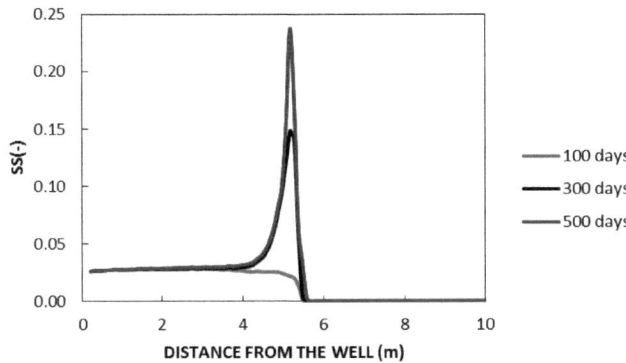

Figure 20: Salt saturation profile for the high permeability middle layer in the layered reservoir at different time snapshots

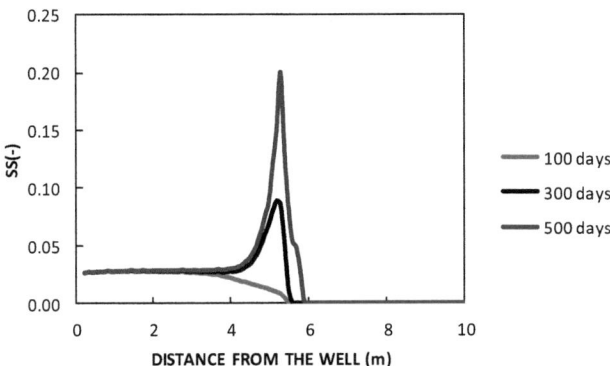

Figure 21: Salt saturation profile for the low permeability bottom layer in the layered reservoir at different time snapshots

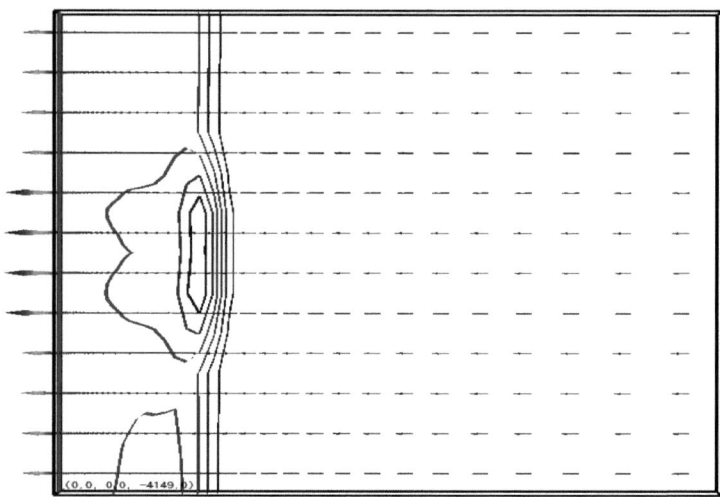

Figure 22: Vector plot of the salt being produced initially in the middle layer before going to the other regions

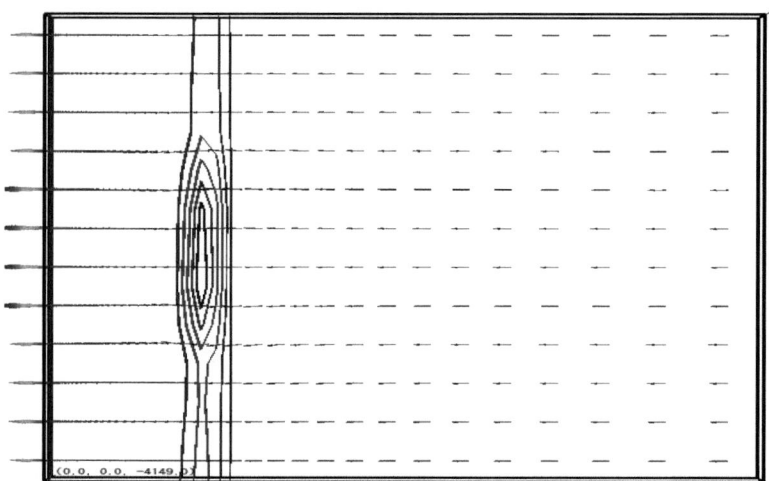

Figure 23: At a later time towards end of simulation the middle layer completely blocks first followed by the bottom layer which is affected by salt

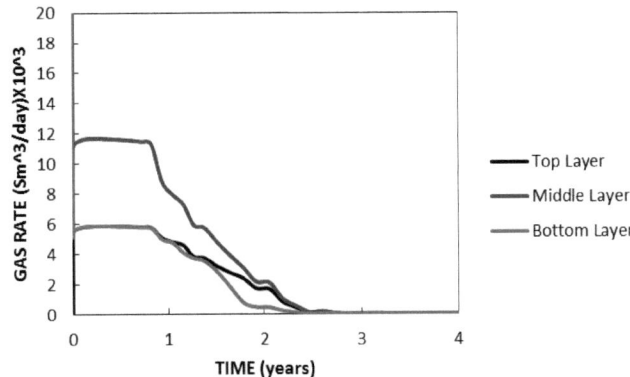

Figure 24: Comparison of the gas productivity rates between the three different layers of the layered model

Influence of Gravity

From figure 24 we concluded that salt precipitation varies in a particular layer with respect to the depth irrespective of homogenous conditions. The only reasonable explanation for the difference in production rates was the influence of gravity. We did another test in the absence of gravity and saw in figure 25 that gravity influences the rate of decline of gas productivity and hence the salt precipitation. In this figure the top and the bottom curves are superimposable with each other which confirm that the model is dependent on the effect of gravity.

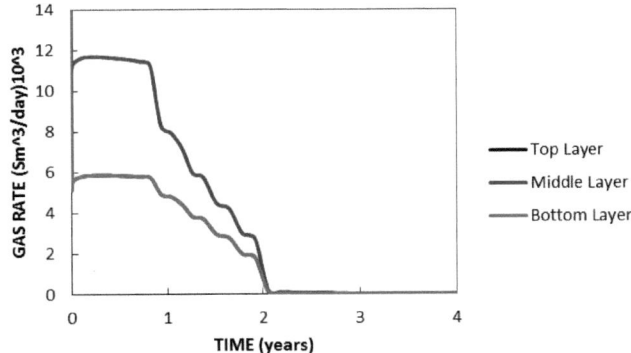

Figure 25: Effect of absence of gravity on the top and the bottom layers of the layered reservoir

41

Effect of Increasing the Permeability Ratio

The final parameter that we tested on the layered reservoir was the alteration of the permeability in the middle layer. Our results show that a change in the permeability ratio does not influence the location of the salt deposition from the well bore. The salt is deposited in a similar manner as in the homogenous reservoir as can be seen in figure 26. However, the amount of salt deposited and the rate of salt deposition is dependent upon the permeability ratio. In the figure we observe that if the permeability of the layer increases the amount of salt saturation will also increase. Applying a permeability factor of 2 the amount of salt deposited in 300 days results in a value 0.05, while applying a factor of 7.5 the amount of salt deposited in 300 days is 0.28. And the well is completely blocked at an earlier time interval. This phenomenon was expected because increasing the permeability increases the flow which in turn increases the rate of evaporation. A higher rate of evaporation results in increased amount of salt deposits at a faster rate as confirmed by the model.

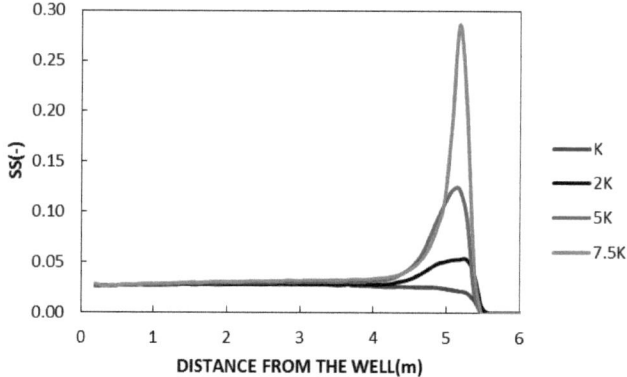

Figure 26: Effect of changing the permeability ratio on the location of salt deposition (300 days)

4.3 General Discussion

From literature it is evident that the salt precipitation problem is a growing one and modelling it is complex in nature. Our radial flow model describes the general location of the salt deposition from the well bore region. We have tried to model the well in such a manner in which we can observe the salt being deposited and the impact it can have on the productivity of gas flow. Overall, we observe that the salt is deposited in a certain regime of 4-5.5m from the well in both the homogenous as well as the heterogeneous versions of the model. However, it is subject to conditions and

certain parameters such as salinity of brine, differential pressure and initial liquid saturation can change the location of the salt as it can deposit further away from the well.

Changing these parameters can speed up the salting phenomena which can reduce the productivity of the well at earlier time intervals and ultimately kill the well. Our results are the first step towards this study in which we have shown the location of salt deposition and its dependency upon certain factors in order to set up a base for future studies regarding this subject.

5. CONCLUSIONS AND RECOMMENDATIONS

Conclusions

From this study we concluded that:

- The model is sensitive to grid refinement
- In both reservoir models (homogenous and layered model) the salt is deposited around 4-5.5 meters from the well.
- The model is sensitive to parameters like initial liquid saturation, salinity content of the brine and a change in differential pressure. The change in liquid saturation effects the location of salt at earlier time intervals around 300 days. As the time progresses the model is not influenced by the initial liquid saturation. However, both the differential pressure and salinity content are sensitive parameters and influence both the rates and the location of the salt significantly as salt seems to deposit further away by increasing the differential pressure and with increased amounts of salt in the brine.
- Presence of heterogeneity in the model does not influence the location of salt deposition but only has an impact on the rate of salt deposition.
- Despite being identical the top and the bottom layers of the layered reservoir behave differently based on their position in the model. Gravity explains the difference in behaviour of gas productivity in these layers.

Recommendations

Our recommendations from this study included:

- This model was made assuming a large cell for the well region. The other approach that can be taken is that a well can be included and the results of both can be compared for a comparative study to observe the location of salt deposition
- We investigated parameters like initial liquid saturation, differential pressure and salinity content with certain permeability relations like the Van Genauchten . The effect of these parameters can be investigated with other in built relations of TOUGH2 like the Corey relations which can give a better insight and for comparative study purposes.

- Our model was made assuming the Bundle of Tubes relationship. Different permeability porosity relations can be used to see if there is any impact on the location of the salt deposition.
- In this study we assumed a certain heterogeneous field (higher permeability surrounded by lower permeability) assuming a certain structure. In reality heterogeneity can be random, so a new model based on a random heterogeneous field can be made to investigate the impact of heterogeneity on salt deposition. The two can then be compared and a further understanding of the subject can be achieved.

APPENDIX

Appendix A: Choosing a Suitable Mesh Size for the Layered Reservoir

In the grid refinement section we had discussed how and why we chose the 540 grid size for our base case. Our simulations for the layered reservoir consisted of the same grid size horizontally but the question arose for choosing a specific number of layers within each layer i.e. the top, middle and bottom. We tried three different cases and the results are shown in figures 27 and 28. Our cases included simulations with 2, 4 and 8 layers within each layer and based on the results we decided to choose 4 layers within each specific layer. The reason we did not choose 2 was it was too course and 8 would take too long making our simulations slow and not optimal. Also there is not much of a difference between the 4 layered model and the 8 layered one as it almost super imposes with each other. Keeping these things in mind we decided to choose 4 individual layers in each main layer of the layered reservoir and are simulations were based on these results.

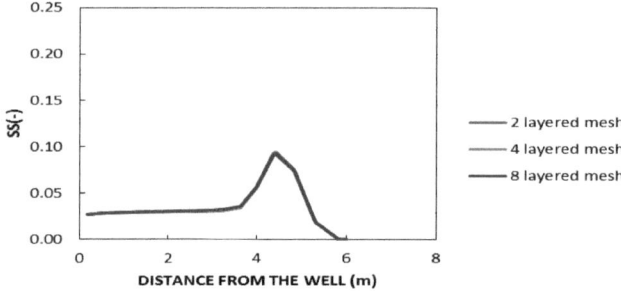

Figure 27: Comparison of the different meshes in the layered model at 300 days

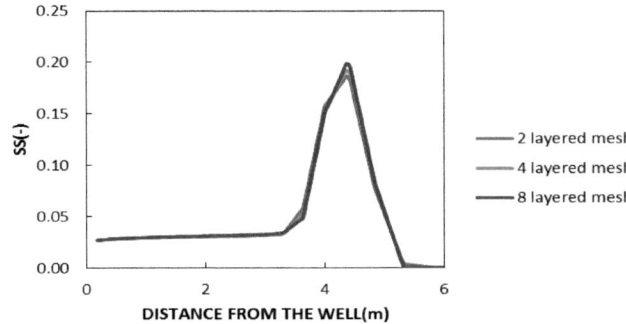

Figure 28: Comparison of the different meshes in the layered model at 450 days

Appendix B: Comparison of TOUGH2 Results with DUMUX

The results that we obtained were from TOUGH2. A similar model was made in DUMUX and the results were compared in order to check the validity of the results for the base case. Figure 29 shows that there is an excellent match in the pressure profile curve while in figure 30 the liquid saturation curve has some differences between TOUGH 2 and DUMUX. We observed that DUMUX had a similar range for the location of the salt deposited (from the liquid saturation curve 4-5.5m) which gave an idea that the salt is depositing in a particular range irrespective of the simulator being used and later studies can be conducted keeping this location in mind which can further provide a better understanding of the subject. In figure 31 the permeability reduction in both the simulators shows that there is a reasonable match between the two where the reduced permeability is achieved at a distance of 4.5m from the well. This confirms that the salt deposits around this range which was one of the goals of this study.

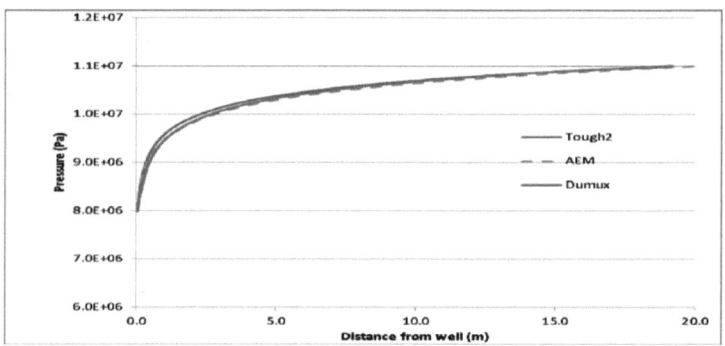

Figure 29: Comparison of the pressure profile curve for TOUGH2 and DUMUX showing an excellent match

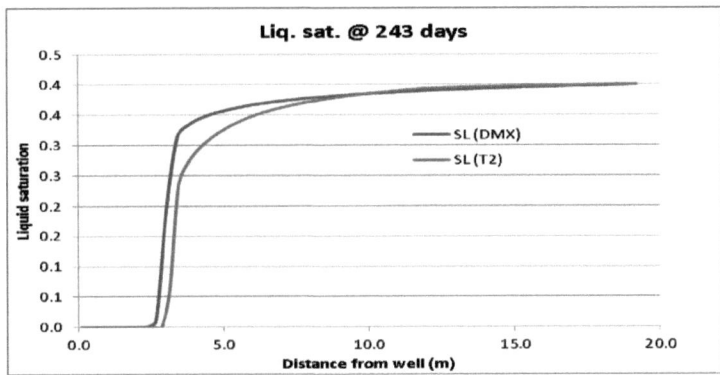

Figure 30: Liquid saturation profile showing a good match between TOUGH2 and DUMUX showing a similar range for the location of the salt deposits with respect to the well

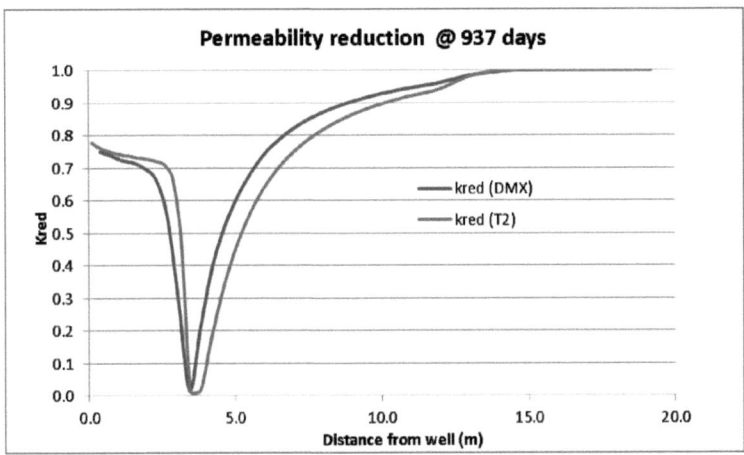

Figure 31: Permeability reduction factor in both the simulators is around 4.5m showing a good match and gives a first idea about the location of salt deposition

Appendix C: Permeability Porosity Relationships in TOUGH2

As mentioned in chapter 3 TOUGH2 has several permeability and porosity relationships already embedded in it. One of these relations is the Verma Pruess relationship which has a couple of variants which can be used. The Bundle of Tubes relationship has been used in our model and is described in equations 2.22-2.24.

Another form of a porosity permeability relationship is a second variant of the Verma Pruess which is known as the Fracture in Series model. The only difference in this relationship is the exponent 2 in the above mentioned equations is replaced by 3 as shown in the following equations:

$$\frac{\kappa}{\kappa_0} = \theta^3 \frac{1 - \Gamma + \Gamma/\omega^3}{1 - \Gamma + \Gamma[\theta/(\theta + \omega - 1)]^3} \qquad (A\text{-}1)$$

Where

$$\theta = \frac{1 - S_S - \Phi_r}{1 - \Phi_r} \qquad (A\text{-}2)$$

$$\omega = 1 + \frac{1/\Gamma}{1/\Phi_r - 1} \qquad (A\text{-}3)$$

In a special case if only straight capillary tubes of uniform radius are considered where both the critical porosity and the fractional length are considered zero equation 2.22 will be as follows:

REFERENCES

1. Kleinitz, W., Kohler, M. and Dietzsch, G.2001. The Precipitation of Salt in Gas Producing Wells.

2. Morin, E., Montel, F. 1995. Accurate predictions for the Production of Vaporized Water.

3. Van Dorp Q. T., Slijkhuis M., and Zitha P.L.J., 2009. Salt Precipitation in Gas Reservoirs.

4. Aquilina, Michael P., Senergy Ltd. Impairment of Gas Well Productivity by Salt Plugging: A Review of Mechanism, Modelling, Monitoring Methods and Remediation Techniques.

5. Zuluaga, E., Munoz, N.I., and Obando, G.A. 2001. An Experimental Study to Evaluate Water Vaporization and Formation Damage Caused by Dry Gas Flow in Porous Media.

6. Kamath, J. and Laroche, C. 2003. Laboratory Based Evaluation of Gas Well Deliverability Loss Caused by Water Blocking.

7. Mahadeven, J., Sharma, M.M., and Yortos, Y.C. 2006. Flow Through Drying of Porous Media.

8. van Genuchten, M.Th. 1980 A Close Form Equation for Predicting the Hydraulic Conductivity of Unsaturated Soils.

9. Tang, R. and Etzion, Y. Comparative Studies on the Water Evaporation Rate From a Wetted Surface and that From a Free Water Surface. (Elsevier(July)2003)

10. Place, Jr, M.C. and Smith, J.T. 1984. An Unusual Case of Salt Plugging in a High Pressure Sour Gas Well.

11. Jasinski, R. Frigo, D. 1997 The Modelling and Prediction of Halite Scale

12. Zuluaga, E., Monslave, J.C, 2003. Water Vaporization in Gas Reservoirs

13. Pruess, K., Oldenburg, C, Moridis, G., 1999. TOUGH2 User's Guide, Version 2.0

14. Battistelli, A., Calore, C., Pruess, K., 1997, The Simulator TOUGH2 /EWASG For Modelling

15. Geothermal Reservoirs with Brines and Non Condensable Gas.

16. Lorenz , S., Muller, W., 2003. Modelling of Halite Formation in Natural Gas Storage Aquifers.

17. Verma, A., Pruess, K. 1988. Thermo-hydrological Conditions and Silica Redistribution Near High Level Nuclear Wastes Emplaced in Saturated Geological Formations.

18. Nicolaides, C. 2014. Salt Precipitation Due to CO_2 Injection into Brine-saturated Heterogeneous Porous Media. Master Student Thesis at TU Delft.

19. Leeuwenburgh, O., Tambach, T., Velthuis, H., Mass, J., Blokland, H. 2011. Salt Precipitation in Gas Reservoirs. (TNO Report).

20. Petrasim User Manual by Thunderhead Engineering.

21. Egberts, P.J.P 2014. Comparison TOUGH2/EWASG and DuMu[x]. Presentation TNO October 2014.

Printed by Books on Demand GmbH, Norderstedt / Germany